YOUR KNOWLEDGE HAS VALUE

- We will publish your bachelor's and
  master's thesis, essays and papers

- Your own eBook and book -
  sold worldwide in all relevant shops

- Earn money with each sale

Upload your text at www.GRIN.com
and publish for free

Bibliographic information published by the German National Library:

The German National Library lists this publication in the National Bibliography; detailed bibliographic data are available on the Internet at http://dnb.dnb.de .

Imprint:

Copyright © 2018 GRIN Verlag
Print and binding: Books on Demand GmbH, Norderstedt Germany
ISBN: 9783668655638

This book at GRIN:

https://www.grin.com/document/413980

Bhavtosh Kikani

Production and Characterization of Bacterial Thermostable Cellulase

GRIN Verlag

Production and Characterization of Bacterial thermostable Cellulase

Author

Bhavtosh A. Kikani

Corresponding Author

Bhavtosh A. Kikani, PhD (Microbiology), ICAR-NET

Assistant Professor

Department of Biological Sciences

PD Patel Institute of Applied Science

Charotar University of Science and Technology

ACKNOWLEDGEMENT

I express my gratitude towards many people who saw me through this book; to all those who provided support, talked things over and assisted in the editing, proofreading and design.

Above all, I thank the Management of Charotar University of Science and Technology for continuous encouragement. In addition, I thank the Principal and Dean, PD Patel Institute of Applied Sciences for the support. I also thank Department of Biological Sciences for the critical comments to improve.

My special gratitude to Prof. S.P. Singh, Professor (Microbiology) and Head, Saurashtra University, Rajkot and Dr. Rajesh K. Sani, Associate Professor, South Dakota Schools of Mines and Technology, USA for continuous support, guidance and timely support.

At this moment, I seek blessings of my parents and in-laws. At the same time, I thank my family members, my wife and my little son who sacrificed more than me to see me achieving the goals.

I also wish to thank GRI PUBLICATIONS for the opportunity.

Last and not least: I beg forgiveness of all those who have been with me and whose names I have failed to mention."

Date: February 22, 2018   **Bhavtosh A. Kikani, PhD**

CONTENTS

INTRODUCTION

Life thrives well in range of adverse environmental conditions, for example extremes of salinity, acidity, alkalinity, temperature or pressure. It always fascinates the scientific fraternity to explore the microbial diversity and phylogeny under these inhospitable habitats. Their possible adaptive measures may provide clues to various evolutionary pathways (Austin, 1988; Herbert and Sharp, 1992; Singh, 2006). Among them, the thermophilic bacteria are common in soil and volcanic habitats with limited species composition. Yet, they possess all the major nutritional categories similar to their mesophilic counterparts. Among, the genus *Bacillus* and related genera are widely distributed in the nature, including thermophilic, psychrophilic, acidophilic, alkaliphilic and halophilic bacteria, which can be able to utilize a wide range of carbon sources for the heterotrophic or autotrophic growth (Claus and Berkeley, 1986; Nazina *et al.*, 2001).

On the other hand, cellulose is the most abundant biomass on Earth, being the primary product of photosynthesis in the terrestrial environment and the most plentiful renewable bioresource. Cellulases, a group of enzymes commonly breaks cellulose, are produced by several microorganisms, mainly by bacteria and fungi. They are inducible enzymes which are synthesized by microorganisms during their growth on cellulosic materials. The complete enzymatic hydrolysis of cellulosic materials needs different types of cellulases: namely endoglucanase (1,4-β-d-glucan-4glucanohydrolase), exocellobiohydrolase (1,4-β-d-Glucan glucohydrolase) and β-glucosidase (β-d-glucoside glucohydrolase). The endoglucanase randomly hydrolyzes β-1,4 bonds in the cellulose molecule, whereas the exocellobiohydrolases in most cases release a cellobiose unit, showing a recurrent reaction from chain extremity. Lastly, the cellobiose is converted to glucose by β-glucosidase.

Bacteria, having quite higher growth rates as compared to fungi contribute significantly in cellulase production. Cellulase yields rely on various factors, mainly inoculums size, pH, temperature, presence of inducers, medium additives, aeration, growth time, and so on. Enormous cellulosic wastes are getting accumulated day by day. Therefore, it is of considerable economic interest to develop the processes for effective treatment and utilization of cellulosic wastes as inexpensive carbon sources.

The cellulose-degrading enzymes can be used, for example, in the formation of washing powders, extraction of fruit and vegetable juices, and starch processing. Cellulases are used in the textile industry for cotton softening and denim finishing; in laundry detergents for colour care, cleaning; in the food industry for mashing; in the pulp and paper industries for drainage improvement and fibre modification, and they are even used for pharmaceutical applications. In nutshell, the cellulose enzymes is commonly used in many industrial applications, whereas the demand for more stable, highly active and specific enzymes will also grow rapidly in near future. Therefore, continuous research for advances in speckled aspects for cellulose production (such as cost, substrate specificity, and specific activity) is desired to achieve improved techno-economic feasibility (Sethi et al., 2013).

In the present study, we intended to isolate potent cellulase producing thermophilic bacteria, followed by analysis of its growth and morphological properties. We optimized its production followed by partially purification of the cellulase by ammonium sulphate fractionation. The study focused on some interesting aspects related to production of the cellulases which would be helpful for its commercialization.

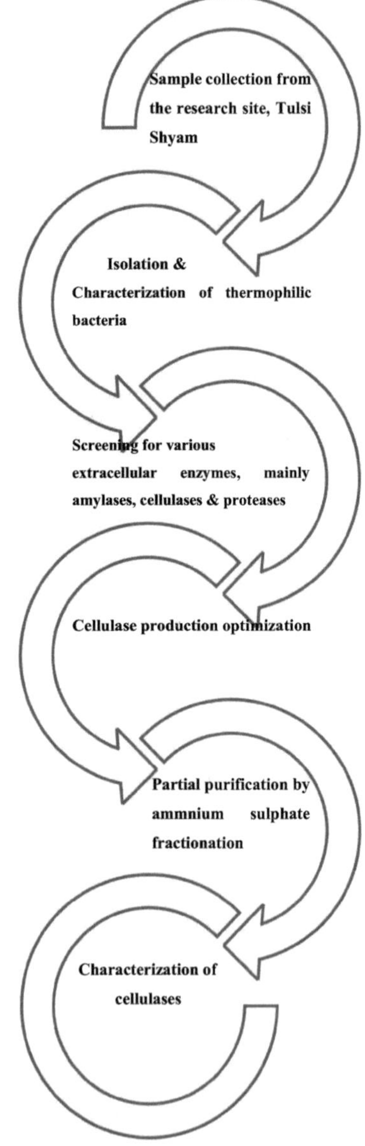

Sample collection from the research site, Tulsi Shyam
Isolation & Characterization of thermophilic bacteria
Screening for various extracellular enzymes, mainly amylases, cellulases & proteases
Cellulase production optimization
Partial purification by ammnium sulphate fractionation
Characterization of cellulases

REVIEW OF LITERATURE

Life at the extreme conditions

The majority of microbial populations, inhabiting in the extreme environments belong to the domains, archaea and bacteria. These habitats are uniquely noticeable due to existence of varieties of extreme conditions, such as salinity, temperature, pH and oxygen concentrations (Stetter, 1998). The microorganisms, optimally flourishing in these unusual environmental conditions, are broadly grouped under the term '*Extremophiles*'. Besides, this microbes are classified based on the nature of the extremity in which they flourish. For instance, psychrophiles grow well in extremely cold environments. Similarly, the microbes which are inherently adapted to very high temperatures are referred to as thermophiles. Further to add, the microorganisms associated to the acidic environments are called acidophiles, whereas those found in the highly alkaline conditions are classified as alkaliphiles. In continuation, many of them uniquely flourish well in presence of two or more type of extreme conditions are known as polyextremophiles. Take an example of haloalkaliphiles, which can grow in the presence of high salt and alkaline conditions. Similar explanation is possible for the thermoalkalitolerant microorganisms. Fascinatingly, the adaptation to thrive in these conditions draws the attention, which may explain their evolution as well (Wiegel and Kevbrin, 2004).

The thermal environments

The most common thermal biotopes are the volcanically and geothermally heated hydrothermal systems, such as the solfataric fields, the natural hot springs and the submarine saline hot vents. The hot solfataric fields consist of one upper layer, containing significant amounts of oxygen, which reflects an ochre color owing to the presence of the ferric ions; while a layer below shows a blackish-blue color, due to the presence of ferrous ions. On the same note, the submarine thermal systems consist of the hot fumaroles, springs, sediments and hot vents with very high temperatures (Stetter, 1998). In addition, the other submarine hydrothermal systems usually contain high concentrations of NaCl, exhibiting a slightly acidic to alkaline pH range between 5 and 8 (Horikoshi, 1998). Interestingly, the thermal environments of the natural hot springs differ very widely with respect to the temperature, flow rate and chemistry of the water (Brock, 1994). The Yellowstone National Park (Wyoming, USA) covers maximum numbers of the hot springs in the World. In addition, the hot springs can also be found at Norris and Mud Vulcano in Italy, Kamchatka in Russia (Wiegel, 1990), along the western coast of India (Saha, 1993), Sao

Michel in Azores, submarine hot springs in Iceland and Mount Grillo at Baia Naples in Italy (Romano *et al.*, 2004). In addition to the above mentioned, other examples of thermal environments include hydrothermal vents at the Guaymas Basin and east Pacific Rise in Mexico. The hydrothermal vents are located in shallow and abyssal depths (Stetter, 1998). These environments possess a unique chemical composition, such as high content of sulfur and hydrogen sulphide, favouring growth of chemolithotrophs.

The biotic communities in the thermal environments

Taxonomically and ecologically, varieties of microorganisms can thrive in the thermal environments. Evidently, microbial life in these habitats represents all the three domains at the high temperatures. Reportedly, taxonomic studies on characterization of members belonging to the bacteria, archaea and eukarya had been conducted in the Yellowstone National Park (Ward, 1998). Specifically, these studies unfolded bacterial diversity, mainly cyanobacteria, anoxygenic phototrophs, aerobic and anaerobic chemoorganotrophs (Hugenholtz *et al.*, 1998; Ward, 1998). In addition, several other unusual thermoacidophiles, such as *Sulfolobus acidocaldarius* were also reported from the Yellowstone National Park (Ward, 1998). Several taxonomic studies were also carried out and reported on the other diverse hot springs. For example, the Mono Lake in California – an alkaline, hypersaline and closed basin was explored by Gorlenko *et al.* (2004). A novel species, *Anaerobranca californiensis* was isolated from the sediment. Another hot spring that has been studied extensively is the Kamchatka at Russia, where *Thermoproteus uzoniensis*, an extremely thermophilic bacterium was isolated (Bonch-Osmolovskaya *et al.*, 1990). Apart from that, numerous thermophilic microorganisms have also been isolated from the chimneys, sediments and the ambient water of the hydrothermal vent fields (Reysenbach *et al.*, 2000).

Thermotoga maritime was isolated from the marine thermal vents at Vulcano (Huber *et al.*, 1986). A similar organism, *Thermotoga neapolitana* was isolated from a submarine thermal vent at Lucrino, Italy (Belkin *et al.*, 1986; Jannasch *et al.*, 1988). Another specific group of thermophiles, including some bacilli, have also been reported from the natural and the artificial high temperature biotopes (McMullan *et al.*, 2004). According to them, the thermophilic bacilli, belonging to the *Bacillus* Genetic Group-5, had been reclassified as the members of the recently named genus *Geobacillus*. Mostly, *Geobacillus* species are widely distributed in the continents, where geothermal areas occur. The geobacilli are also isolated from the shallow marine hot springs and from the deep-sea hydrothermal vents as well. Maugeri *et al.* (2002) previously described the isolation of three novel halotolerant and thermophilic *Geobacillus* strains from three separate shallow marine vents of the Eolian Islands, Italy. Besides, high temperature oil fields also yielded strains of *Geobacillus*, where two novel species, namely *G. subterraneus* and *G. uzenensis* were isolated from the Uzen oil-field in Kazakhstan (Nazina *et al.*, 2001; Nazina *et al.*, 2004). In addition, *Geobacillus* species were isolated from the temperate soils (McMullan *et al.*, 2004) and other artificial hot environments, such as hot water pipelines, heat exchangers, waste treatment plants, burning coal refuse piles and bioremediation biopiles (Maugeri *et al.*, 2002; Obojska *et al.*, 2002). On the other hand, archaea is the least understood in terms of the diversity, physiology and ecological panorama of the three primary phylogenetic domains. Although many species of Crenarchaeota (Woese *et al.*, 1990) have been isolated, they constitute a relatively tight-knit cluster of lineages in the phylogenetic analyses of the nucleic acid sequences. It seemed possible that this limited diversity is merely apparent and reflects only a failure to culture the organisms and not their absence. It obviates need for cultivation and identification of the organisms. This may have a definite impact on the concepts of the phylogenetic organization of archaea. Furthermore, use of molecular phylogenetic approaches in the microbial ecology has revolutionized the view of the microbial diversity and has led to the proposal of a new kingdom within the Archaea, namely the Korarchaeota (Reysenbach *et al.*, 2000). The report consisted on the occurrence of another member of the archaeal group and a deeply rooted bacterial sequence from a thermal spring in the Yellowstone National Park. The phylotype is a lineage within the *Aquificales*. The *In-situ* hybridization with bacteria-specific and *Aquificales*-specific fluorescent oligonucleotide probes indicated that the bacterial populations dominated the community and contributed significantly to the biogeochemical cycling within the community (Reysenbach *et al.*, 2000).

2.2.2 The adaptations against the elevated temperatures

Modifications to the protein structure for survival at the higher temperature have been extensively reviewed (Fields, 2001), with most research into the field of thermophilic enzymes (Kikani *et al.*, 2010, Singh *et al.*, 2013). As thermophilic microorganisms cannot shield their internal environment from the external temperature, the cellular components have get adapted at the elevated temperatures. The studies of the extremophilic proteins have revealed no structural motifs, covalent modifications or additional amino acids that would explain the ability of the proteins to function in the extreme conditions. The analysis of the structural data had shown that a redistribution of the some of the forces ensures the stability in the mesophilic environments and the changes in the protein-solvent interaction are sufficient to maintain the structural integrity at the high temperatures as well. In turn, the approach allowed the rapid modification in the enzyme stability against the environmental changes, by simply modifying the concentration of solutes, which help in the adaptation to a new thermal niche (Fields, 2001).

Besides, high temperatures increase the fluidity of the cellular membrane, which thereby lead to 'leaky' membranes and therefore consequently loss of the functioning of the membrane proteins. However, in the thermophilic bacteria, the fatty acids are more saturated and longer than the fatty acids in the mesophiles. The polar liquids are also enriched in the carbohydrates, containing a greater proportion of the methyl branched fatty acid chains (Mermelstein and Zeikus, 1998). The archaeal cytoplasmic membrane contains unique ether-linked lipids, which are very resistant to the high temperatures and do not degrade (De Rosa *et al.*, 1994). They are also resistant to the mechanical degradation and very high salt concentrations, making these lipids more competent for the cellular membranes of the thermophiles than the eubacterial estertype lipids (Vossenberg *et al.*, 1998). Evidently, the tetraether membrane lipids were reported in a thermoacidophilic euryarchaeota, *Candidatus 'Aciduliprofundum boonei'* (Schouten *et al.*, 2008).

In addition, the cell wall of the extremely thermophilic archaea is reported to be composed of an 'S-layer' structure – a simple, regular two-dimensional lattice of the glycoproteins that covers the cell surface and leaves no periplasmic space. The glycoprotein layers are highly resistant to the mechanical and chemical degradation, while they spontaneously re-assemble due to very strong subunit interactions (Herbert and Sharp, 1993). The cell wall of an alkaliphilic *Bacillus* contained an acidic teichurono peptide polymer, which served as a barrier to ionic flux and played a role in the pH homeostasis (Kitada *et al.*, 2000). Besides, cell wall of several alkaliphilic microbes also contain a large amount of the acidic amino acids. The acidic charges may act as charged membranes, reducing the pH on the cell surface between 8 and 9, allowing the cell to maintain a neutral internal pH (Horikoshi, 1998). Also, all hyperthermophiles contain the enzyme, reverse gyrases to induce the positive super-coiling of DNA that enhances its thermal stability (Forterre, 1996). In addition, the thermophilic microorganisms are reported to have proteins which are thermostable and resist denaturation and proteolysis (Kumar and Nussinov, 2001). The proteins and enzymes of the thermophilic microorganisms can also adapt to high temperatures by increased electrostatic, disulphide and hydrophobic interactions (Ladenstein and Ren, 2006; Pebone *et al.*, 2008). Additionally, certain specialized proteins, known as 'chaperons', are produced by these organisms, which help to refold the proteins to their native form and restore their functions (Laksanalamai and Robb, 2001; Singh *et al.*, 2010). Besides, certain thermophilic enzymes are stabilized by various conformational changes (Fitter, 2003). However, presence of certain metals, inorganic salts and substrate molecules are also reported to impart the thermostability (Vieille and Zeikus, 2001).

The enzymes of the thermophilic microorganisms can be used as models to understand the basis of thermostability. In addition, studies of the thermophilic organisms and their proteins has provided important insight into the mechanisms of the protein folding. Therefore, understanding how thermophilic proteins have evolved to be stable can yield information about the functional modulation of the folding landscapes. It has been shown, in studies comparing homologous proteins from the thermophilic and mesophilic organisms, those proteins from the thermophilic organisms have a lower change in the heat capacity upon unfolding. It is thought that this is due to residual structure in the unfolded state of the protein from the thermophilic organism.

2.3 The Biocatalytic potentials of thermophilic bacteria

As a consequence of the growth at high temperatures and unique macromolecular properties, the thermophilic bacteria also possess high metabolic rates, physically and chemically stable enzymes and lower growth but higher end product yields. Moreover, the reactions at very higher temperatures have benefits of the decreased viscosity and the enhanced diffusion coefficient of substrates, favoring equilibrium displacement in the endothermic reactions (Kumar and Swati, 2001). Thus, many possibilities for industrial processes have emerged with the thermostable enzymes (Haki and Rashit, 2003).

2.3.1 Cellulases

Biotechnology of cellulases and hemicellulases began in early 1980s, first in animal feed followed by food applications (Chesson, 1987). Subsequently, these enzymes were used in the textile, laundry as well as in the pulp and paper industries (Wong and Saddler, 1992). Currently, several commercial enzyme producers are marketing tailor-made enzyme preparations suitable for biotechnology (Bhat and Bhat, 1997).

MATERIAL AND METHODS

The present section focuses on the screening of these bacteria, with respect to their capabilities to produce various extracellular enzymes. Besides, various physicochemical environmental conditions were also optimized for the production of cellulases. At the end, cellulose was partially purified and characterized accordingly.

The Research site, Tulasi Shyam – A natural thermal habitat

The research site, Tulasi Shyam is a set of the natural hot spring reservoirs, located in the middle of the Gir Forest in the Gujarat State, India (Latitude: 21.051; Longitude: 71.025). Noticeably, the proposed site also has a religious significance. Various soil samples and water samples were collected from the site.

Characterization of the samples

Various physicochemical properties of the soil and water samples were measured. The parameters for the water sample included colour, temperature and pH values. The parameters for the soil samples included colour, pH, soil structure, soil texture and soil consistence (Marx *et al.*, 1999; McSweeney and Grunwald, 1999).

Isolation of the thermophilic bacteria

The samples were serially diluted in the sterile distilled water. From the diluted samples, 5% (v/v) was inoculated into the modified thermophilic medium containing 0.7% (w/v) peptone, 0.5% (w/v) , malt extract and 0.5% (w/v) NaCl, along with 1% (w/v) CMC powder at different pH from 6-8. The flasks were incubated at 50°C for 24–48h. Thereafter, 0.1ml of the enriched culture was inoculated on the modified thermophilic agar plates with 5% (w/v) agar containing the same ingredients as mentioned above at pH 6-8. The inoculated plates were incubated at the temperatures between 37 and 70°C for 24–48h. The colonies were subsequently streaked repeatedly till pure colonies were obtained.

Screening for the various extracellular enzymes

In order to reveal the biocatalytic potentials, the bacterial isolates were individually spotted on different screening medium plates.

Preparation of activated culture

The thermophilic bacterial cultures were inoculated in the modified medium, containing 0.5% (w/v) peptone, 0.5% (w/v) yeast extract, 0.3% (w/v) malt extract, 0.2% (w/v) $(NH_4)_2SO_4$, 0.5% (w/v) NaCl, 0.07% K_2HPO_4, 0.03% KH_2PO_4, 0.05% $MgSO_4.7H_2O$ and 0.01% $CaCl_2$. The cultures were incubated at their optimum growth conditions for 16-18h. The young actively growing culture, also commonly known as the activated culture was used further.

Detection of the Amylases

The activated bacterial cultures were spotted on the starch agar plates, containing 2% (w/v) starch, 0.5% (w/v) yeast extract, 0.5% (w/v) peptone, 0.5% (w/v) NaCl and 5% (w/v) agar powder, with different pH values ranging between 6 and 9. Simultaneously to check the effects of temperature, the plates were incubated at different temperatures: 45, 50, 55 and 60°C for 48h. The plates were flooded with the iodine solution (Hi Media, India). Formation of a clear halo zone surrounding the colony against dark blue background confirmed production of extracellular amylase.

Detection of the Proteases

To screen for the production of the proteases, the activated thermophilic bacterial cultures were spotted on the gelatin agar plates containing, 3% (w/v) gelatin, 1% (w/v) peptone, 0.5% (w/v) NaCl and 5% (w/v) agar powder, with different pH values ranging between 6 and 9. Subsequently, the plates were incubated at different temperatures: 45, 50, 55 and 60°C for 48-72h. The production of extracellular protease was confirmed by clear zone surrounding the colony upon addition of Frazier's reagent, containing 5g $HgCl_2$ and 20ml concentrated HCl in 100ml distilled water.

Detection of the Cellulases

Similarly, the attempts to detect the extracellular cellulases were made on the Dubo's cellulose agar plates, containing 1% (w/v) cellulose, 1% (w/v) $NaNO_3$, 0.1% (w/v) K_2HPO_4, 0.01% (w/v) $FeSO_4.7H_2O$, 0.05% (w/v) KCl, 0.05% (w/v) $MgSO_4.7H_2O$, 0.5% (w/v) NaCl and 5% (w/v) agar powder, with different pH values ranging between 6 and 9. The plates were incubated at different temperatures: 45, 50, 55 and 60°C for 48h. The Dubo's cellulose agar plates were flooded with the iodine solution (Hi Media, India), where the appearance of the clear zone surrounding the colony confirmed the production of extracellular cellulases.

Detection of the Lipase

The lipolytic activity of the organisms were investigated, using the tributyrin agar plates, containing 1% (v/v) tributyrin, 1% (w/v) yeast extract, 0.5% (w/v) peptone, 0.5% (w/v) NaCl and 5% (w/v) agar powder, with varying pH values between 6 and 9. The activated cultures were spotted individually on the plates. The plates were further incubated at range of temperatures: 45, 50, 55 and 60°C for 48h. The appearance of the clear zone surrounded the colony confirmed the production of lipase.

Optimization of the physicochemical factors affecting the production of the extracellular cellulases

Assay to measure the cellulase activity

Cellulase was assayed by measuring the reducing sugar released during the reaction, using carboxymethyl cellulose (CMC) as the substrate. The reaction contained 0.5ml the enzyme and 0.5ml of 1% (w/v) CMC, dissolved in 20mM phosphate buffer, pH 7. The reaction mixture was incubated at the optimum temperature for the enzyme activity for 20 min. The reaction was stopped by adding 1ml of dinitrosalicylic acid solution. The dinitrosalicylic acid solution (100ml) consisted of 1g 3, 5-dinitrosalicylic acid, 30g potassium sodium tartarate and 20ml 2N NaOH solution. Further, the reaction mixture was heated in the boiling water for 10 min. The reaction mixture was cooled in ice and diluted with 8ml distilled water, prior to the spectrophotometric measurement of the absorbance at the wavelength, 540nm. Besides, one unit of cellulase was defined as the amount of enzyme liberating 1μmol of glucose per minute under the assay conditions, using glucose (100–1000g) as a standard.

Effect of incubation temperature and medium pH on the cellulase production

The activated cultures were individually inoculated in the CMC broth containing, 1% (w/v) CMC, 0.5% (w/v) yeast extract, 0.5% (w/v) peptone, 0.5% (w/v) malt extract and 0.5% (w/v) NaCl. Both, the medium pH [range: 6, 7, 8 and 9] and the incubation temperature [range: 37, 45, 50, 55 and 60°C] were kept variable. The flasks were incubated for 120h. Bacterial growth and cellulase activity in each flask were monitored regularly at the regular interval of 24h. The sets were performed in the triplicates.

Effect of CMC concentration on the cellulase production

The activated cultures were individually inoculated in the starch broth with aforementioned composition. However, the CMC concentration [range: 0.5, 1, 1.5 and 2% (w/v)] was kept variable, while the medium pH and the incubation temperature were constant, i.e. the optimum medium pH and incubation temperature for the respective bacterial isolate. The flasks were incubated for 120h. Bacterial growth and cellulase activity in each flask were monitored regularly at the regular interval of 24h. The sets were performed in the triplicates.

Effect of carbon sources on the cellulase production

Apart from CMC, other simple carbon sources including glucose, lactose, maltose, sucrose and galactose were supplemented in the concentration of 1% (w/v) into the CMC broth. The medium pH, CMC concentration and incubation temperature were kept constant at the optimum value, as described earlier. The carbon sources were kept variable in each flask. The flasks were incubated for 120h. Bacterial growth and cellulase activity in each flask were monitored regularly at the regular interval of 24h. The sets were performed in the triplicates.

Effect of nitrogen sources on the cellulase production

In order to check the effect of various nitrogen sources on the production of cellulase, different nitrogen sources including the organic nitrogen sources, such as gelatin, yeast extract, malt extract and peptone and the inorganic nitrogen sources, such as ammonium sulphate and ammonium nitrate were supplemented in the CMC broth. Besides, these nitrogen sources were also supplemented in various combinations to investigate any effect on the production of cellulases. The medium pH, incubation temperature, CMC concentration and the carbon sources and their respective concentrations were kept constant, i.e. optimum for the respective bacterial isolate. The nitrogen source was kept variable in each flask. The flasks were incubated for 120h. Bacterial growth and cellulase activity in each flask were monitored regularly at the regular interval of 24h. The sets were performed in the triplicates.

Partial purification by ammonium sulphate fractionation

The crude cellulase solution was adjusted to 0-50% (w/v) and 50-70% (w/v) ammonium sulfate fractionation. The recovered precipitates were suspended in the minimum volume of 0.1M phosphate buffer, pH 7. Later, the fractions were dialyzed under same buffer for 24-48 h. This was followed by estimation of total proteins and total cellulase activity to determine specific activity. This was also used to calculate fold purification and purity (%) at a particular purification step. The protein content was measured by the Bradford method using bovine serum albumin (10-100μg/ml) as a standard.

Characterization of the Cellulase

The temperature and pH optima

Cellulase activity was determined at various temperatures from 25 to 90°C. The activity was also measured in the following buffers: 20mM acetate buffer, pH 5, 20mM phosphate buffer, pH 6–7, 20mM tris–HCl buffer, pH 8, 20mM glycine–NaOH buffer, pH 9 and 20mM NaOH–Borax buffer, pH 10.

The temperature and pH stability of cellulase

The cellulase thermostability experiments were carried out between 25°C to 90°C for 24 h. The pH stability was carried out in various buffers, pH 6–10 for 24 h. The residual activity can be calculated by considering activity at 0 Min as 100%.

As thermophilic bacteria can be a potential candidate in the present day biotechnological industries, the bacteria were screened for their enzymatic potentials. Moreover, better upstream processing always leads to the best product, in terms of the quality and the quantity. Therefore, various physicochemical factors affecting the production of cellulases were optimized.

The hot spring reservoirs of Tulasi Shyam, Gujarat (India)

This site is located in Gujarat, India. From ecological point of view, it is the only natural thermal habitat in the Saurashtra region. Enormous microbial diversity was well evident in soil and water samples, which can be explored to capture their respective biotechnological significance, mainly the extracellular enzymes.

Properties of the soil and water samples

The physicochemical properties of the soil and water samples were described in Table-1. The pH of the soil sample was slightly acidic to neutral. The granular structure of the soil indicated that it was rich in organic matters, with good porosity and easy water air exchange. The water sample was transparent in nature. Temperature was 50°C because of natural volcanic and geothermal processes. The pH was neutral.

Table-1. Properties of the soil and water samples

Parameters	Properties
Soil samples	
pH	**6.7**
Structure	**Granular**
Color	**Black**
Texture	**Loam**
Water samples	
Transparency	**Semi-transparent**
Colour	**Greenish**
Temperature	**50°C**
pH	**7**

The colony characteristics of the thermophilic bacteria

In total, eight thermophilic bacterial isolates were obtained. The colony characteristics were described in Table-2.

Table-2. Colony characteristics of the thermophilic bacteria

Isolate	Colony characteristics					
	Shape	**Margin**	**Texture**	**Elevation**	**Opacity**	**Pigmentation**
Isolates from the soil samples						
TS-1	Round	Regular	Mucoid	Flat	Opaque	White
TS-2	Round	Regular	Dry	Raised	Opaque	Off white
TS-3	Round	Irregular	Mucoid	Raised	Opaque	Pale yellow
TS-4	Flowery	Irregular	Mucoid	Flat	Opaque	Pale yellow
Isolates from the water samples						
WS-1	Round	Regular	Dry	Raised	Opaque	Off white
WS-2	flowery	Irregular	Dry	Flat	Opaque	White
WS-3	Round	Regular	Dry	Flat	Opaque	Off white
WS-4	Flowery	Irregular	Mucoid	Raised	Opaque	Off white

Growth and morphological properties of the thermophilic isoates

The thermophilic bacterium was isolated by enrichment culture technique. Growth was observed between 45-70°C and pH 6-9. The Gram reactions and morphologies of bacteria were described in Table-4. Absence of growth at 37°C indicated true thermophilic nature of the isolate.

Table-4. Growth and morphological properties

Bacterial isolate	Optimum growth pH (Range)	Optimum growth Temperature (Range)	Gram Reaction and Morphology
TS -1	6 (6-7)	50°C (45-70°C)	Gram Positive very short thin rods, arranged singly

TS-2	7 (6-9)	50°C (45-70°C)	Gram Positive short thin rods, arranged chain
TS-3	8 (7-9)	50°C (45-70°C)	Gram Positive short thin rods, arranged singly
TS-4	8 (6-9)	50°C (45-70°C)	Gram Positive long thin rods, arranged singly
WS-1	8 (7-8)	50°C (45-70°C)	Gram Positive, thin cocobacilli, arranged in chain
WS-2	9 (7-9)	50°C (45-70°C)	Gram Positive short thin rods, arranged singly
WS-3	9 (8-9)	50°C (45-70°C)	Gram Positive short thin cocobacilli, arranged singly
WS-4	9 (6-9)	50°C (45-70°C)	Gram Positive long thin rods, arranged chain

Screening for the various extracellular enzymes

Thermophilic bacteria, dwelling in the natural hot spring reservoir, Tulasi Shyam in the Gujarat State, India were screened for the secretion of the extracellular enzymes, mainly amylase, protease, cellulase and lipase [Table 5].

Table-5. Screening of the thermophilic bacterial isolates for various enzymes

Isolate	Zone Ratio (Z/C)			
	AMYLASE	CELLULASE	PROTEASE	Lipase
TS-1	1.2	0	0	0
TS-2	1.4	1.2	0	0
TS-3	1.6	0	0	0
TS-4	2	2.1	0	0
WS-1	1.5	0	0	0
WS-2	1	1.2	0	0
WS-3	0.8	0	0	0
WS-4	1	1.8	0	0

Evidently, not a single bacterium could secrete proteases and lipase. However, many of them could secrete amylases and cellulases, together [Table-5].

Identification of *Anoxybacillus rupiensis* TS-4 (GenBank: KU360725)

The studied thermophilic bacterium was isolated by enrichment culture technique from the soil sample from the natural hot spring reservoirs of Tulasi Shyam in Gujarat, India. The organism was long rod displaying Gram positive character. Moreover, absence of growth at 37°C indicated the thermophilic nature of the isolate. Based on 16S rRNA gene sequence analysis, the isolate was designated as *Anoxybacillus rupiensis* TS-4 (GenBank: KU360725).

Biochemical properties of the thermophilic bacterial isolate, TS-4

The biochemical properties of the thermophilic bacterial isolate, TS-4 is listed as below (Table-6).

Table-6. Biochemical test of thermophilic bacteria, *Anoxybacillus rupiensis* TS-4

BIOCHEMICAL TEST	*Anoxybacillus rupiensis* TS-4
Gelatin hydrolysis	Negative
Starch hydrolysis	Positive
Cellulose hydrolysis	Positive
Lipid hydrolysis	Negative
Catalase test	Positive
Oxidase test	Negative
Sugar fermentation test	
Glucose	Positive
Sucrose	Negative
Lactose	Negative
Maltose	Positive
Fructose	Negative
Mannitol	Negative
Galactose	Negative
Phenyl alanine deamination	Negative
Citrate utilization	Negative

TSI test	
Acid formation in slant	Positive
Acid formation in butt	Positive
Gas production	Positive
H_2S Production	Negative
Indole formation	Negative
Urea degradation	Positive
Methyl Red test	Negative
Voges-Proskauer test	Positive
Nitrate reduction test	Positive
Ammonia production	Positive

Optimization of the cellulase production

Various physicochemical factors affecting production of cellulases were studied extensively.

The obtained trends are as follows:

Effect of the Incubation Temperature and medium pH on the cellulase production

The incubation temperature and the medium pH have immense effects on the microbial growth, which in turn reflects in the production of the extracellular enzymes. Therefore, the effects of incubation temperature and medium pH on the bacterial growth and the cellulase production were investigated. The bacterium produced cellulases within the temperature range between 45 and 60°C. Yet, the optimum temperature was 50°C [Fig. 3].

With regard to the optimum medium pH, the trends revealed that the thermophilic bacteria optimally produced the cellulases at neutral medium pH. Yet, some alkalitolerance was apparent [Fig. 3].

Fig. 3 The temperature and medium pH optimization for the production of the cellulase by thermophilic bacterium TS-6; pH 7 (--), pH 8 (--) and pH 9 (--)

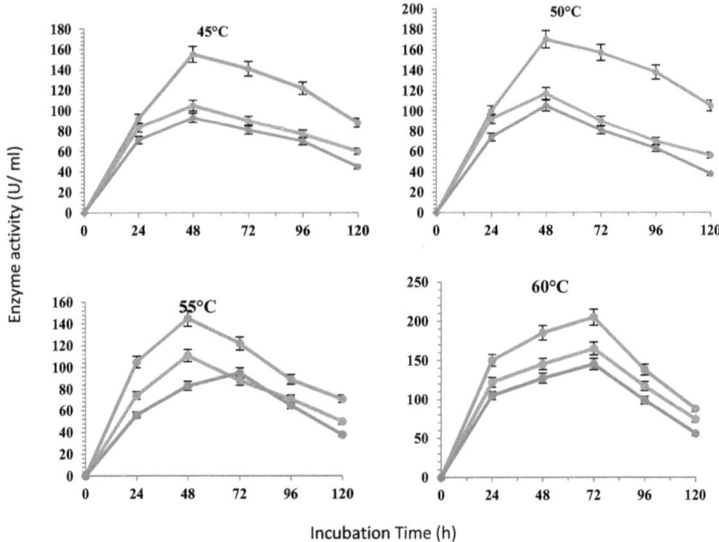

Effect of various carbon sources on the cellulase production

Subsequent to the deduction of the optimum incubation temperature, the medium pH and the required CMC concentration, the effects on the cellulase production due to supplementation of various carbon sources were investigated. The carbon sources supplemented were glucose, lactose, maltose, sucrose and galactose, in 1% (w/v) concentration. Interestingly, presence of any of the simple sugar resulted in a remarkable inhibition of the cellulase production [Fig. 4], albeit the microbial growth enhanced astonishingly. The results supported the general theory of 'the glucose effect' or in other words, 'the catabolite repression' as the bacteria utilized the simple sugars first, which was reflected by the enhanced microbial growth. However, it led to repression of the cellulase production. Therefore, present results revealed that the supplementation of the simple sugars to the production medium was not suitable for the optimal cellulase production.

Effect of different nitrogen sources on the cellulase production

The nitrogenous sources are one of the significant medium ingredients. It promotes microbial growth and production of enzymes as well. The growth of each bacterium affected greatly on the type of the nitrogen source supplied, i.e. organic and inorganic nitrogen sources or the nitrogen sources in the combination, which were in turn reflected in the production of the cellulases.

Comparatively, microbial growth and cellulase production together were favoured on addition of the combinational nitrogen sources. As revealed, the organic nitrogen source alone did produced cellulases satisfactorily but the inorganic nitrogen source, as a sole nitrogen source led to very poor microbial growth and the enzyme production as well [Fig. 5].

Fig. 4 Effect of various simple sugars on the production of the cellulose

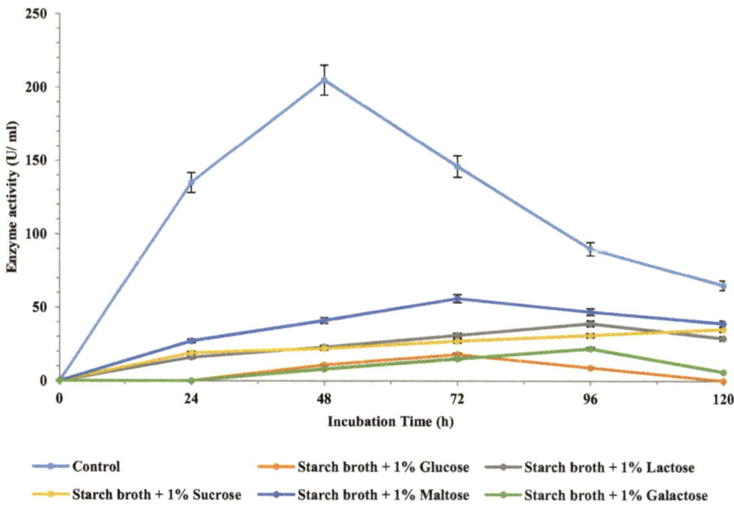

Fig. 5 Effect of various nitrogen sources on the production of the cellulose

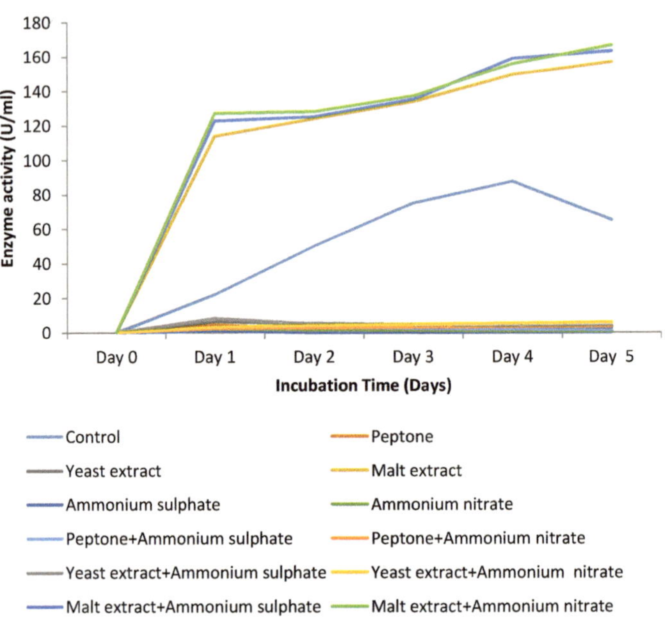

Partial purification of thermostable cellulases

The purification table (Table-7) indicates fold purification (8.22) and yield (39%) along with step wise increase in the specific activity.

Table-7. Purification Table indicating fold purification and percent yield

Purification step	Volume (ml)	Total activity (Units)	Total protein (mg)	Specific activity (Units/ mg)	Purification fold	Yield (%)
Crude	1000	205000	400	512.5	-	100
0-50% fraction	40	18000	25	720	1.40	8.78
51-70% fraction	40	72000	20	3600	7.02	35.12
Dialyzed	40	80000	19	4210.5	8.22	39.02

ENZYME CHARACTERIZATION

Temperature and pH profile

Each and every enzyme has its unique temperature and pH optimum, at which its catalytic activity can be observed at its best. The studied cellulase was broadly active at range of temperature and pH with the optimum values to be 70°C and 7, respectively [Fig. 6 (a)-(b)].

Fig. 6 (a) Temperature optimization of partially purified cellulase

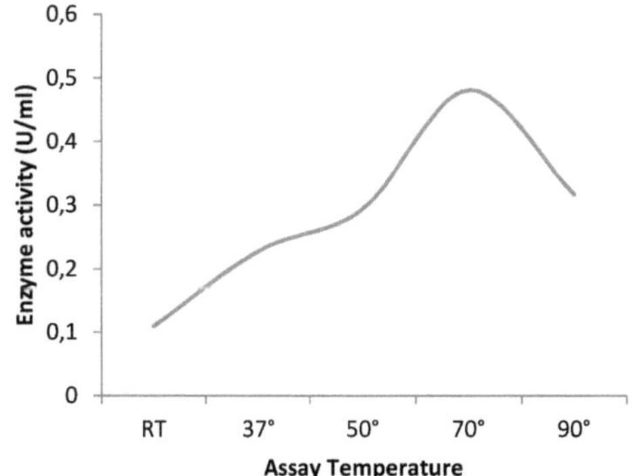

Fig. 6 (b) pH optimization of partially purified cellulase

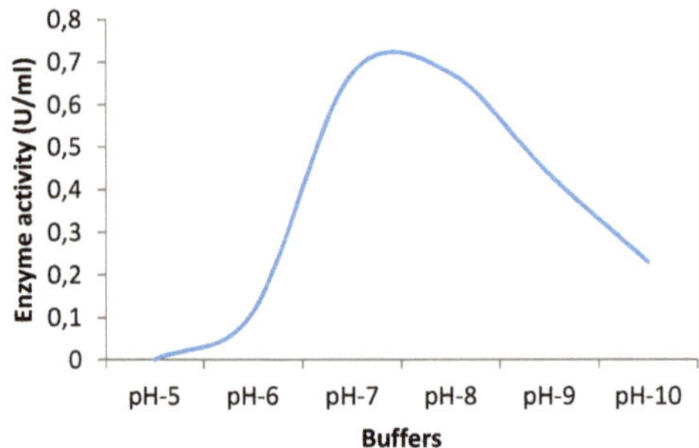

Temperature and pH stability of cellulases

The cellulase of *A. rupiensis* TS-4 was very thermostable. The enzyme was stable even after 24 h at the elevated temperatures. [Fig. 7 (a)]. With regard to the pH stability, the cellulase was stable in various buffers with different pH at 50°C. The enzyme was optimally stable in 20mM phosphate buffer, pH 7. Besides, the enzyme remained stable wide range of pH values: pH 5-10 for 24h, exhibiting its potential commercial significance [Fig. 7 (b)].

Fig. 7 (a) Temperature stability of partially purified cellulases

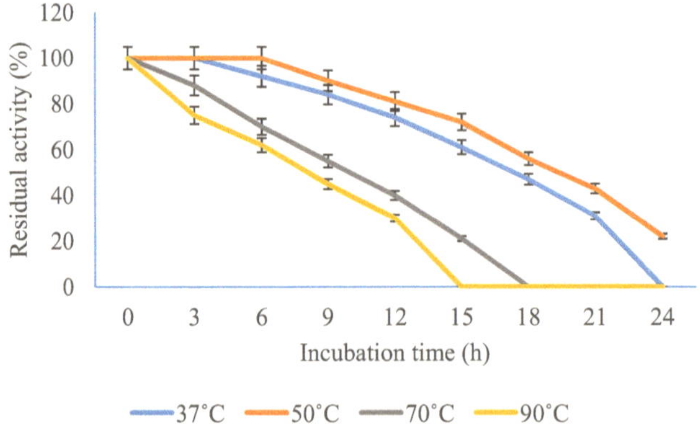

Fig. 7 (b) pH stability of partially purified cellulases

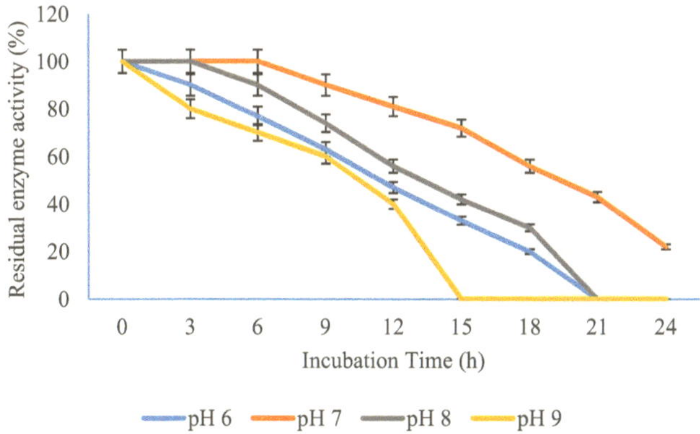

DISCUSSION

Tulsi Shyam is the only natural thermal habitat in the Saurashtra region, Gujarat. Various thermophilic bacteria were isolated, identified and screened for amylases and cellulases. The physicochemical conditions affecting cellulase production were optimized. Later, it was partially purified, characterized and explored further for degradation of cellulosic wastes.

Some physical factors, *viz.,* temperature and pH have substantial effects on growth, enzyme production, enzyme activity and stability. Both, growth and enzyme production increased with increasing temperature within the range of 45°C to 60°C with an optimum of 50°C. Interestingly, the isolates could not grow well at mesophilic range, which is a clear cut indication of their thermophilic nature. The results also strongly reflected that the optimum temperature for growth as well as cellulase production was almost same. A similar trend has been reported for *Bacillus* species (Toedoro and Martine, 2000). The results also suggested that isolate produced cellulase optimally at neutral to alkaline pH and that the higher enzyme production at this pH was a result of increased cell growth. The same was the case where neutral pH was found to be optimal for cellulase production as also reported in *Bacillus* sp. (Acharya and Chaudhary, 2012), *B. subtilis* (Sethi et al., 2013), *Geobacillus pallidus* (Baharuddin et al., 2010).

Along with the physical factors, certain medium ingredients may also influence the growth and enzyme production. The cellulase synthesis by several microorganisms has been correlated to the presence or absence of various amino acids and complex nitrogenous sources in the culture medium. Indeed, the addition of malt extract, peptone and other inorganic nitrogen source to the liquid medium shortened the lag period and increased both the growth of the cell and the enzyme synthesis. Therefore, the result suggests that malt extract, peptone and ammonium sulfate favored the growth and production of thermostable amylases by the organism studied. The same is the case with *Bacillus* sp. (Toedoro and Martine, 2000). It has been reported that the synthesis of carbohydrate degrading enzymes in most species of the genus *Bacillus* is subject to catabolic repression by readily metabolizable substrates such as glucose. Our results are in good agreement with these findings, as addition of glucose (1%) to the culture diminished greatly the synthesis of cellulase. These results are exactly similar with the findings from hyperthermophilic archaeon *Sulfolobus solfataricus*, where production of amylases was repressed by glucose and glucose prevented amylases gene expression and not merely secretion of preformed enzyme (Haseltine *et al.,* 1996). Addition of calcium and yeast extract enhanced microbial growth and enzyme production, whereas glucose at 1% level showed a strong repression (Asgher et al., 2007). Certain trace metal ions can also play a substantial role.

Besides, the cellulases or the cellulolytic organisms can be potentially used in the bioethanol fermentation, which can be used as a biofuel. The present trends indicated that either the enzyme alone or the bacteria, itself hydrolyzed corn cob very efficiently. Similar trends were reported by Zambare et al., 2011, where they used bioagro wastes to produce bioethanol using cellulases treatment. Similar trends were also evident by Bhalla et al., 2014, Binod et al., 2010 and Cardona et al., 2010.

Thus, the cellulase was produced optimally in the CMC broth, pH 7 containing 0.5% peptone, 0.5% malt extract, 0.2% ammonium sulphate, 0.2% ammonium nitrate and 0.2% NaCl at 50°C by a thermophilic bacterial isolate, *A. beppuensis* TS-4. Besides, the partially purified cellulase and whole cell could hydrolyze cellulosic waste very remarkably, indicating its future applications in production of bioethanol, i.e. biofuels.

SUMMARY

The thermophilic bacteria were isolated from Tulsi Shyam, Gujarat. The biochemical and morphological properties were checked. None of the isolate could grow at mesophilic range, confirming their true thermophilic nature. The cellulase production was optimized by Placket Burman method, followed by central composite design method. The three factorial optimization designs revealed the optimum temperature, medium pH and CMC concentration to be 50°C, 7 and 1% (w/v), respectively. Later, it was found that simple sugars and trace metal ions could not enhance cellulase production. However, presence of malt extract, ammonium nitrate and ammonium nitrate could enhance the cellulase production significantly. The enzyme was partially purified and ultimately used for the hydrolysis of the corn cob. The decreased viscosity and increased reducing sugars indicated efficient hydrolysis of the cellulolytic wastes.

REFERENCES

1. Acharya, S., Chaudhary, A., (2012). Optimization of fermentation conditions for cellulases production by *Bacillus licheniformis* MVS1 and *Bacillus* sp. MVS3 isolated from Indian hot spring. *Braz. Archives of Biol. Technol.* 55 (4): 497-503.

2. Asgher, M., Asad, M.J., Rahman, S.U., Legge, R.L. (2007) A thermostable α-amylase from a moderately thermophilic *Bacillus subtilis* strain for starch processing. *J. Food Eng.* 79: 950-955.

3. Austin, B. (1988) Methods in Aquatic Bacteriology. A Wiley Inter science Publication, USA.

4. Baharuddin, A.S., Abd, R., Lim, S., Mohd, N.A., Suraini, A.A., Abdul, R., Md, S., Mohd, A.H., Sakai, K. and Shirai, Y. (2010). Isolation and characterization of thermophilic cellulase-producing bacteria from empty fruit bunches-palm oil mill effluent compost. *American J. Appl. Sci.* 7 (1): 56-62.

5. Belkin, S., Wirsen, C.O., Jannasch, H.W. (1986) A new sulfur-reducing extremely thermophilic eubacterium from a submarine thermal vent. *Appl. Environ. Microbiol.* 51: 1180-1185.

6. Bhalla, A., Bischoff, K.M., Uppugundla, N., Balan, V. and Sani, R.K. (2014) Novel thermostable endo-xylanase cloned and expressed from bacterium Geobacillus sp. WSUCF1. *Bioresource technol.* 165: 314-318.

7. Bhat, M.K. and Bhat, S. (1997) Cellulose degrading enzymes and their potential industrial applications. *Biotechnol. Adv.* 15 (3-4): 583-620.

8. Binod, P., Sindhu, R., Singhania, R.R., Vikram, S., Devi, L., Nagalakshmi, S., Kurien, N., Sukumaran, R.K. and Pandey, A. (2010) Bioethanol production from rice straw: an overview. *Bioresource technol.* 101 (13): 4767-4774.

9. Bonch-Osmolovskaya, E.A., Miroshnichenko, M.L., Kostrikina, N.A., Chernych, N.A., Zavarzin, G.A. (1990) *Thermoproteus uzoniensis* sp. nov., a new extremely thermophilic archaebacterium from Kamchatka continental hot springs. *Arch. Microbiol.* 154: 556-559.

10. Brock, T.D. (1994). Life at high temperatures. Yellowstone Association for Natural Science, History and Education Inc., Yellowstone National Park, Wyoming, pp. 34.

11. Cardona, C.A., Quintero, J.A. and Paz, I.C. (2010) Production of bioethanol from sugarcane bagasse: status and perspectives. *Bioresource technol.* 101 (13): 4754-4766.

12. Chesson, A. (1987) Supplementary enzymes to improve the utilization of pigs and poultry diets. In: Haresign, W., Cole, D.J.A. (Ed.) Recent advances in animal nutrition. Butterworths publications, London, pp. 71–89.

13. Claus, D., Berkeley, RC.W. (1986) The Genus *Bacillus*. In: Sneath P.W. (Ed.) Bergey's Manual of Systematic Bacteriology, Volume 2. Wilkins, Baltimore, pp. 1105-1139.

14. De Rosa, M., Morana, A., Riccio, A., Gambacorta, A., Trincone, A., Incani, O. (1994) Lipids of the archaea: a new tool for bioelectronics. *Journal of Biosensors and Bioelectronics.* 9: 669–675.

15. Fields, P.A. (2001) Protein function at thermal extremes: balancing stability and flexibility. *Comparative Biochem. Physiol.* 129: 417-431.

16. Fitter, J. (2003) A measure of conformational Entropy change during thermal protein unfolding using neutron spectroscopy. *Biophysical J.* 84 (6): 3924-3930.

17. Forterre, P., Bergerat, A., Lopezgarcia, P. (1996) The unique DNA topology and DNA topoisomerases of hyperthermophilic Archaea. *FEMS Microbiol. Rev.* 18: 237-248.

18. Gorlenko, V., Tsapin, A., Namsaraev, Z., Teal, T., Tourova, T., Engler, D., Mielke, R., Nealson, K. (2004) *Anaerobranca californiensis* sp. nov., an anaerobic alkalithermophilic, fermentative bacterium isolated from a hot spring on Mono Lake. *Int. J. Syst. Evol. Microbiol.* 54: 739-743.

19. Haki, G.D., Rakshit, S.K. (2003) Developments in industrially important thermostable enzymes. *Bioresource Technol.* 89: 17-34.

20. Haseltine, C., Rolfsmeier, M., Blum, P. (1996) The glucose effect and regulation of α-amylase synthesis in the hyperthermophilic archaeon *Sulfolobus solfataricus*. *J. Bact.* 178: 945-950.

21. Herbert, R.A., Sharp, RJ. (1992) Molecular Biology and Biotechnology of Extremophiles. Chapman and Hall, New York, USA.

22. Horikoshi, K. (1998) Alkalophiles. In: Horikoshi, K., Grant, W. (Ed.) Extremophiles: Microbial Life in Extreme Environments. Wiley-Liss, Inc. New York, pp. 35-56.

23. Huber, R., Langworthy, T.A., Koning, H., Thomm, M., Woese, C.R., Sleytr, U.B., Stetter, K.O. (1986) *Thermotoga maritime* sp. nov. represents a new genus of unique extremely thermophilic eubacteria growing up to 90° C. *Arch. Microbiol.* 144: 324-333.

24. Hugenholtz, P., Goebel, B.M., Pace, N.R. (1998) Impact of Culture Independent Studies on the Emerging phylogenetic view of bacterial diversity. *J. Bact.* 180: 4765-4774.

25. Jannasch, H.W., Huber, R., Belkin, S., Stetter, K.O. (1988) *Thermotoga neapolitana* sp. nov. of the extremely thermophilic eubacterial genus *Thermotoga*. *Arch. Microbiol.* 150: 103-104.

26. Kikani, B.A. Singh, S.P. (2011) Single step purification and characterization of a thermostable and calcium independent α-amylase from *Bacillus amyloliquifaciens* TSWK1-1 isolated from Tulsi Shyam hot spring reservoir, Gujarat (India). *Intern. J. Biol. Macromol.* 48 (4): 676-681.

27. Kikani, B.A., Shukla, R.J. and Singh, S.P. (2010) Biocatalytic potential of thermophilic bacteria and actinomycetes. In: Mendez-Vilas, A. (ed.) Current Research, Technology and Education Topics in Applied Microbiology and Microbial Biotechnology, Vol. 2, Formatex Research Center, Badajoz, Spain, pp. 1000-1007.

28. Kitada, M., Kosono, S., Kudo, T. (2000) The Na^+/H^+ antiporter of alkaliphilic *Bacillus* sp. *Extremophiles.* 4: 253-258.

29. Kumar, H.D., Swati, S. (2001) Modern Concepts of Microbiology, Second revised edition. Vikas Publishing House Private Limited, New Delhi (India).

30. Kumar, S., Nussinov, R. (2001) How do thermophilic proteins deal with heat? *Cell. Mol. Life Sci.* 58: 1216–1233.

31. Ladenstein, R., Ren, B. (2006) Protein disulfides and protein disulfide oxidoreductases in hyperthermophiles. *Fed. European Biochem. Soc. J.* 273 (18): 4170-4185.

32. Laksanalamai, P., Robb, F.T. (2004) Small heat shock proteins from extremophiles. *Extremophiles.* 8: 1-11.

33. Marx, E.S., Hart, J., Stevens, R.G. (1999) Soil Test Interpretation Guide, Oregon State University, USA.

34. Maugeri, T.L., Gugliando, C., Caccamo, D., Stackerbrant, E. (2002) Three novel halotolerant and thermophilic *Geobacillus* strains from shallow marine vents. *Syst. Appl. Microbiol.* 25: 450-455.

35. McMullan, C., Rahman, T.J., Banat, I.M., Ternan, N.G., Marchant, R. (2004) Habitat, applications and genomics of the aerobic, thermophilic genus *Geobacillus*. *Biochem. Soc. Trans.* 32: 214–217.

36. McSweeney, K., Grunwald, S. (1999) Soil Morphology, Classification and Mapping. University of Wisconsin-Madison, USA.

37. Mermelstein, L.D., Zeikus, J.G. (1998) Anaerobic non-methanogenic extremophiles. In: Horikoshi, K., Grant, W. (Eds.). Extremophiles: Microbial Life in Extreme Environments. Wiley-Liss, Inc. New York, pp. 71-93.

38. Nazina, T.N., Lebedeva, E.V., Poltaraus, A.B., Tourova, T.P., Grigoryan, A.A., Sokolova, D.S., Lysenko, A.M., Osipov, G.A. (2004) *Geobacillus gargensis* sp. nov. a novel thermophile from a hot spring, and the reclassification of *Bacillus vulcani* as *Geobacillus vulcani* comb. nov. *Int. J. Syst. and Evol. Microbiol.* 54: 2019-2024.

39. Nazina, T.N., Tourova, T.P., Poltaraus, A.B., Novikova, E.V., Grigoyan, A.A., Nvanova, A.E., Lysenko, A.M., Petrunyaka, V.V., Grigoyan, G.A., Ivanova, A.E. Taxonomic study of aerobic thermophilic bacilli: descriptions of *Geobacillus subterraneus* gen. nov. sp. nov. and *Geobacillus uzenensis* sp. nov. from petroleum reservoirs and transfer of *Bacillus stearothermophilus*, *Bacillus thermoglucosidasius* and *Bacillus thermodinitrificans* to *Geobacillus* as the new combinations *G. stearothermophilus*, *G. thermocatenulatus*, *G. thermoleoverans*, *G. kaustophilus*, *G. thermoglucosidasius* and *G. thermodinitrificans*. *Int. J. Syst. and Evol. Microbiol.* 2001, 51: 433-446.

40. Obojska, A., Ternan, N., Lejczak, B., Kafarski, P., McMullan, G. (2002) Organophosphonate utilization by the thermophile *Geobacillus caldoxylosilyticus* T20. *Appl. Environ. Microbiol.* 68: 2081-2084.

41. Pebone, E., Limauro, D., Bartolucci, S. (2008) The machinery for oxidative protein folding in Thermophiles. *Antioxid Redox Signal.* 10 (1): 157-169.

42. Reysenbach, A.L., Ehringer, M., Hershberger, K. (2000) Microbial diversity at 83°C in Calcite Springs, Yellowstone National Park: another environment where the *Aquifiacales* and "Korarchaeota" coexist. *Extremophiles.* 4: 61-67.

43. Romano, I., Lama, L., Moriello, V. S., Poli, A., Gambacorta, A., Nicholause, B. (2004) Isolation of a new thermohalophilic *Thermus thermophilus* strain from hot spring, able to grow on a renewable source of polysaccharide. *Biotechnol. Lett.* 1: 45-49.

44. Saha S. K. (1993) Limnology of thermal springs. Narendra Publishing House, New Delhi, pp. 176.

45. Schouten, S., Baas, M., Hopmans, E.C., Reysenbach, A.L., Damste, J.S.S. (2008) Tetraether membrane lipids of Candidatus *"Aciduliprofundum boonei"*, a cultivated obligate thermoacidophilic euryarchaeote from deep-sea hydrothermal vents. *Extremophiles*. 12: 119-124.

46. Sethi, S., Datta, A., Gupta, B.L., Gupta, S. (2013) Optimization of cellulase production from bacteria isolated from soil, *ISRN Biotechnol*. 2013: 1-7.

47. Singh, S.P. (2006) Extreme Environments and Extremophiles, In: National Science Digital Library (CSIR): E- Book, Environmental Microbiology. CSIR, India, pp. 1-35.

48. Singh, S.P., Shukla, R.J. and Kikani, B.A., 2013. Molecular diversity and biotechnological relevance of thermophilic actinobacteria. In: Satyanarayana, T., Littlechild, J., Kawarabayasi, Y. (Eds.) Thermophiles in Environmental and Industrial Biotechnology, Springer Publication, UK, pp. 459-479.

49. Stetter, K. (1998) Hyperthermophiles: Isolation, Classification and Properties. In: Horikoshi, K., Grant, W. (Eds.). Extremophiles: Microbial Life in Extreme Environments. Wiley-Liss, Inc. New York, pp. 1-24.

50. Toedoro, C.E.D., Martins, M.L.L. (2000) Culture Conditions for the production of thermostable amylase by *Bacillus* sp. *Braz. J. Microbiol.* 31:298-302.

51. Vieille, C., Zeikus, G. (2001) Hyperthermophilic enzymes: sources, uses, and molecular mechanisms for thermostability. *Microbiol. Mol. Biol. Rev.* 65 (1): 1-43.

52. Vossenberg, J.L.C., Driessen, A.J.M., Konings, W.N. (1998) The essence of being extremophilic: The role of the unique archaeal membrane lipids. *Extremophiles*. 2: 163-170.

53. Ward, D.M. (1998). Microbiology in Yellowstone National Park. In: ASM News. The News Magazine of the ASM. 64: 141-146.

54. Wiegel, J. (1990) Temperature spans for growth: a hypothesis and discussion. *FEMS Microbiol. Rev.* 75, 155-170.

55. Wiegel, J., Kevbrin, V.V. (2004) Alkalithermophiles. *Biochemical society transactions*. 32: 193-198.

56. Woese, C.R., Kandler, O., Wheelis, M.L. (1990) Towards a natural system of organisms: proposals for the domains archaea, bacteria and eucarya. *Proc. Natl. Acad. Sci. USA*.87: 4576-4579.

57. Zambare, V.P., Bhalla, A., Muthukumarappan, K., Sani, R.K. and Christopher, L.P. (2011) Bioprocessing of agricultural residues to ethanol utilizing a cellulolytic extremophile. *Extremophiles*. 15 (5), 611.

YOUR KNOWLEDGE HAS VALUE

- We will publish your bachelor's and
  master's thesis, essays and papers

- Your own eBook and book -
  sold worldwide in all relevant shops

- Earn money with each sale

Upload your text at www.GRIN.com
and publish for free